PETER J. RE

Director, Butser Ancient I

ANCIENT FARMING

SHIRE ARCHAEOLOGY

Cover photograph
A prehistoric harvest: a general view of one of the research fields bearing a rich crop of emmer wheat and a plethora of arable weeds at the Butser Ancient Farm Research Project near Petersfield, Hampshire.

British Library Cataloguing in Publication Data available

Published by
SHIRE PUBLICATIONS LTD
Cromwell House, Church Street, Princes Risborough,
Aylesbury, Bucks HP17 9AJ, UK

Series Editor: James Dyer

ISBN 0 85263 876 0

First published 1987

Set in 11 point Times and printed in Great Britain by
C. I. Thomas & Sons (Haverfordwest) Ltd,
Press Buildings, Merlins Bridge, Haverfordwest, Dyfed.

Contents

List of illustrations

Acknowledgements

Since this book is a brief and partial distillation of more than twenty years of empirical research I owe a debt of gratitude to professional and amateur colleagues, organisations, institutions and universities far too numerous to list in detail. However, I would like to thank particularly Mrs Perry, the secretary of the Ancient Farm, for typing and retyping the manuscript, Mr A. Wyman, the assistant director of the Ancient Farm, for his invaluable help in preparing photographs and commenting upon the text, and finally Mr J. Dyer, the editor of Shire Archaeology, for helpful comments and contributions. Naturally the opinions and theories and any errors are the responsibility of the author.

1
Introduction

Prehistory is conventionally divided into three major periods, the stone age, the bronze age and the iron age, the name of each period indicating the prime raw material used for the production of tools and weapons. Similarly we assign crude time zones to these periods, the stone age concluding about 2000 BC, the bronze age about 800-700 BC and the iron age at the time of the Roman conquest in AD 43. The iron age continues until the industrial revolution in technological terms but documented history takes precedence. These divisions represent a comfortable and orderly approach to the prehistoric period and allow for multiple subdivisions and regional variations which both clarify and obfuscate the issues. Without written records our knowledge is entirely dependent upon the surviving material evidence recovered by archaeological excavation, a factor which has led directly to the nomenclature of the different periods. There is little doubt that the people themselves did not share this consciousness of their time position or even recognise their material resources as significant markers. Their initial problems were survival and increased control of their destiny. In simple terms prehistory is the story of agriculture from the first farmers of the neolithic, the late stone age, who began to carve a living from a dominating landscape, to the farmers of the late iron age, who had succeeded in total domination and exploitation of their environment. By the late iron age society was extremely complex, layered and capable of analysis into production industry and service industry with rural and urban settlements.

The objective of this book is to explore how and why agriculture first established itself and then succeeded in such a remarkable manner. While the normal divisions of time will be regularly referred to in order to give a context to the development of agriculture, only the growth of agriculture will be dealt with, from its beginnings with the felling of trees and initial land clearance up to its establishment as the successful economic basis of iron age society. In effect, agriculture is the common theme not only of prehistory but also of the history of Britain virtually to the present day. Some would claim that it still is the fundamental economy of the country.

Because we are relying on the flimsiest of evidence the picture presented can never be entirely accurate but the urgent need is to explain how the landscape of the iron age, which shows man's total domination of the countryside, came about. Similarly we need to explain the import-export industry of the later iron age referred to by the Romans and corroborated by archaeological excavation. Grain and leather, products of arable and pastoral farming, were exported to the continent. The implications of this are that there must have been a regular and organised surplus production with a full infrastructure of service industries processing, transporting and ferrying the surplus under trading conditions.

The classical texts have been known and in a sense totally disregarded until relatively recently because the archaeological evidence was not available to support them. The conventional view of the iron age has been, and still is in some accounts, that of a society struggling to survive and hanging on until the gift of civilisation was made by the Romans in AD 43. By the same token the preceding periods of the bronze and stone ages were comparatively worse. Happily this view has been completely changed by archaeological excavations at sites like Danebury in Hampshire, Fengate near Peterborough, Maiden Castle and Little Woodbury in Dorset. Throughout the British Isles so much more has been discovered of all periods that the assessment has changed completely. The Romans came into contact with a vibrant, successful and populous society in Britain. Indeed, the underlying motives for the conquest as opposed to the overt political reasons could be determined to be economic. The fact that the conquest was relatively unopposed would sustain a theory that the British felt that they would themselves profit from being part of a bigger economic community. Sustaining these lines of argument has been the work of Butser Ancient Farm Research Project in Hampshire, where experiments have been carried out to assess both the agricultural and the domestic economy of the iron age and early Roman period. The results dramatically underline the potential success of that agricultural system as we presently understand it.

2
The nature of the evidence

In order to reconstruct the remote past (the further back in time the more sparse is the material evidence), we are strictly at the mercy of specific indicators: pollen grains, plant seeds carbonised by accident, site excavations, tools and fragments of tools, phytoliths and plant gloss, bones, occasionally waterlogged deposits — the list is pitifully short. However, given this evidence we first need to evaluate the elements and determine the weight of reliance that can be placed upon each category.

Pollens

Pollen grains are the male reproductive elements of the plant world. Microscopic in size and unique to their species, they are dispersed either by wind or by insects or other animals. Many fail to reach their goal and are lost to the reproductive process. Of these the wind-dispersed pollens are the most useful in that ultimately they drift to earth, a proportion falling on to lakes and rivers and finding their way into the sediment layers beneath the water. The critical factor from our point of view is their survival, recovery and identification. Conditions for their survival are quite specific: acidic and anaerobic. Peat bogs are ideal sources for pollen evidence since as the bog grows, often by predictable amounts, so layers are sealed and preserved. By extracting vertical cores through such bogs and lake sediments and analysing the pollen grains present it is possible to build up a picture of the plants, particularly including trees, which were in the region of the bog. Further, given the ability to date organic material by carbon-14 dating techniques, it is possible to describe the general vegetation of particular periods and sequences of periods. It is necessary to make adjustments to the pollen evidence giving due weight to the more and less prolific pollen producers. However, it is still important to remember that such evidence is at best extremely partial. Having achieved a list of plant species from the pollen evidence, it is also possible to draw conclusions about the general climate since particular species require particular climatic conditions to flourish. Indeed, pollen evidence is used to argue the beginning of farming when man began to denude the landscape of trees in order to create fields. Significant reductions of tree pollens in the sample cores without climatic change indicators are used as evidence of this first phase of agriculture.

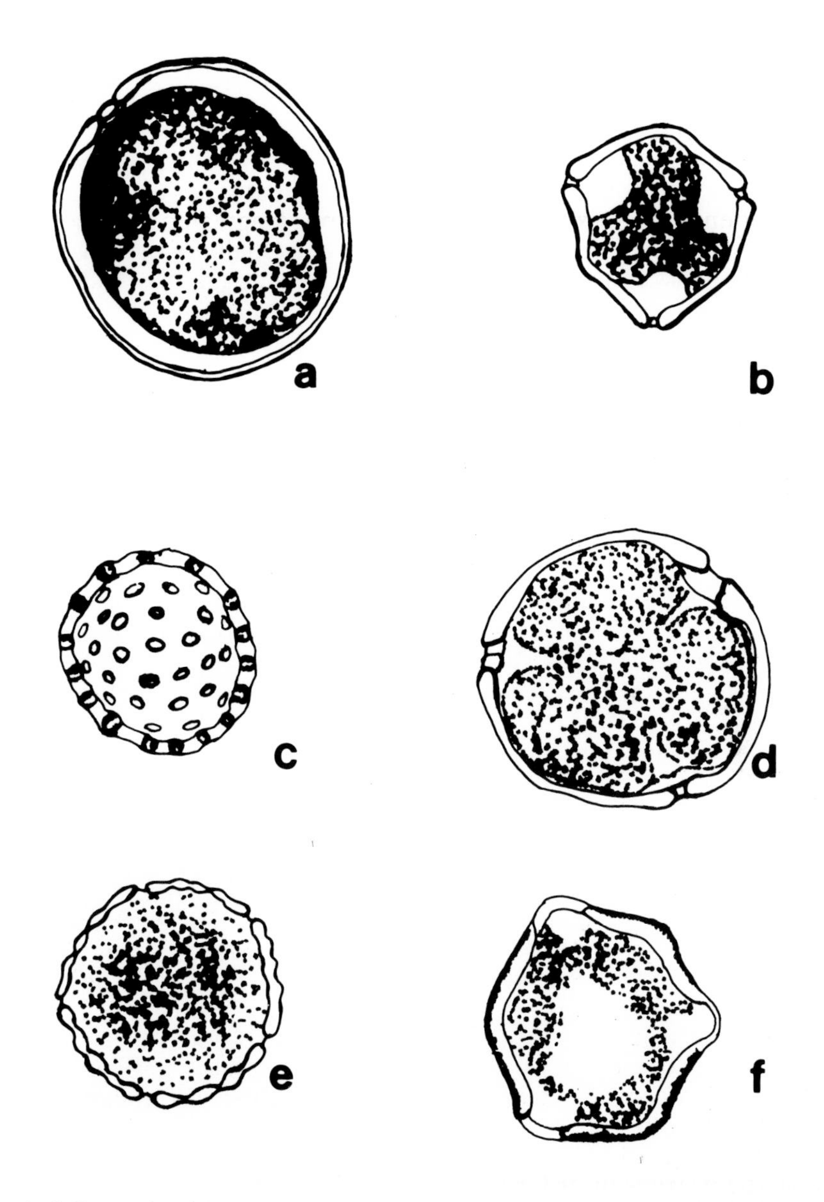

1. Pollen grains (all at a magnification of one thousand times): (a) grass *(Phleum pratense)*; (b) hazel *(Corylus avellana)*; (c) fat-hen *(Chenopodium album)*; (d) beech *(Fagus sylvatica)*; (e) elm *(Ulmus* sp.*)*; (f) oak *(Quercus robur)*.

2. Carbonised seed (left). Occasionally a complete ear of wheat survives like this example of emmer (*Triticum dicoccum*). A latex mould (below) shows the positive impression of a spikelet of spelt (*Triticum spelta*) from a prehistoric pot.

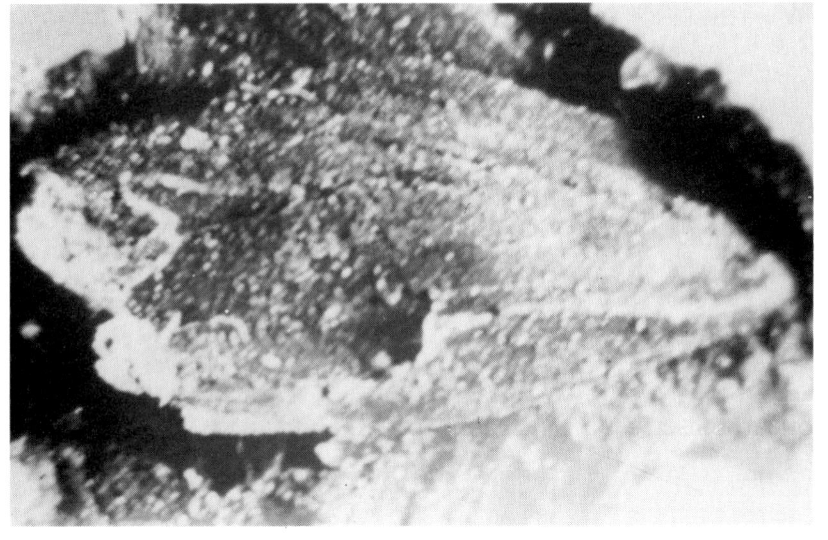

The decline in abundance of elm-tree pollen in the neolithic is often quoted in this instance although this decline could have been caused by a disease akin to the modern Dutch elm disease. From the agricultural point of view pollen evidence is useful since the pollens of the majority of trees and grasses, a subtribe of which are the cereals, are all wind-dispersed. There are, however, problems because of the acidic nature of the survival deposits. Many cereals will not grow on very acid soils and, consequently, their pollens are unlikely to be positively identified in peat-bog samples.

Carbonised seed

In the excavation of settlements of all prehistoric periods one of the most significant discoveries is that of plant seeds which have been burned and turned into charcoal. Improvement in excavation techniques allied to careful sieving of deposits has increased this source of evidence dramatically. Because the seeds have been carbonised they have largely retained their shape and therefore their species is often identifiable. The same is true of wood which has been turned into charcoal. Indeed, charcoal fragments are perhaps the most abundant archaeological indicators of all. However, it must be remembered that not only is the recovery of this type of evidence a mere sample, but by necessity it is a skewed sample. Virtually without exception the finds come from the excavation of settlement sites, which means that in the case of trees and agricultural plants they have been transported into the settlement from the neighbourhood. Thereafter this harvested material is subjected to at least one process before it is burned, allowing chance survival as potential carbonised evidence. This chance survival is so fraught with variables that it is virtually impossible to ascribe specific functions which provide carbonised evidence. Theories abound about this aspect of evidence survival. For example, carbonised grains might be the by-product of drying cereals before processing or storage. The theory, however, begs the question whether it is necessary to dry cereals in the first place.

Another potential source of carbonised seed has been tested empirically with remarkable results. It seems reasonable to suppose that rubbish disposal by bonfire would provide quantities of carbonised seed. The rubbish from thatching straw, the missed ears of wheat at harvest and the arable weed undergrowth cut with the straw, all of which have to be cleaned from the straw bundles before they are used in the thatch, could well have been burned up in a bonfire. Simulations of this process with known quantities of seed input show a regular sixty per cent survival of seeds. Significantly the number of carbonised seeds recovered from sites can be counted in tens and hundreds rather than thousands yet a field yielding approximately a ton to the acre of a prehistoric cereal type like emmer wheat contains 36 million seeds plus or minus five million. The proportions recovered are, therefore, hardly statistically significant beyond a presence and absence list. Even this listing has to take into account the partiality of the harvesting process which will not necessarily gather a fully representative sample of all the plants in the field. This is not to deny the value of presence and absence lists, rather it is to temper the approach to this kind of evidence with great caution.

In contrast there are occasions when the carbonised seed can be used to enhance insight into a function. The best example is those seeds recovered from grain storage pits which bear the traces of the functions of storage and, therefore, may provide indications of preparation before storage and demonstrate the plant community which passes through the preparation process. Unfortunately this evidence is limited in one sense but important in another since the presence and absence listing allows for hypotheses to be raised from comparative modern studies.

Further evidence is provided by impressions of seeds fired into pottery. Frequently potsherds are found with negative impressions in their surface. By taking simple casts from these 'moulds' it has proved possible to identify different plants, especially cereals. It seems that these impressions were purely accidental and not the result of deliberate decoration. However, the same happenstance argument prevails for these as well as the carbonised seed with the added observation that the potters were decidedly careless.

Bones

The survival of bone evidence depends much upon soil type and the nature of the bone itself. Nonetheless from the point of view of agriculture it is critically important. From the bones recovered from a settlement site the palaeozoologist is able not only to identify species but occasionally, given sufficient quantities of material, to build up detailed pictures of the specific animals. By comparison with the bone structures of relic species like the Moufflon and Soay sheep it has proved possible to provide close parallels to the early farm livestock. An additional bonus is the discovery of butchery marks, proving that the animals were kept as food resources. However, bone evidence is even more partial than seed evidence since there are so many reasons why it should not survive or why the collections that do survive are not representative of a total livestock holding. Not least of the variables is the well attested presence of dogs from the stone age onwards. The chances of survival of chicken and other bird bones are remote, of sheep and goat bones slightly less so. The average dog would leave few remains of such bones and their distribution would seem remarkably random to the human if not to the canine mind.

From the evidence which has survived it has proved possible to construct an overall if incomplete picture. For the later iron age there are a few references by the classical writers to farm livestock but these also are not exhaustive and refer mostly to south-east England.

Also it is necessary to view the documentary evidence with great care. The observations made and recorded are not usually corroborated and are unlikely to be accurate. Even more suspicious are those reports which quote original sources which have been lost. Finally nearly all of the documentary material is imbued with political motivation. Nonetheless it does have value provided it is used with care.

Tools

From all periods of prehistory the survival and recognition of tools is critically important. Agricultural implements were fashioned from stone, bone, wood, bronze and iron. Our primary problem is the identification of agricultural tools, assigning to them particular functions and attempting to discover how they may have worked. It is unlikely that we can ever be absolutely certain that our interpretation is accurate. For example, sickles, or tools which look like sickles, have been recovered from all prehistoric periods. In the stone age these are usually composite tools made up of a number of manufactured flint flakes set into a curved wooden handle. Examination of the cutting edges and adjacent areas shows evidence of plant gloss and phytolith deposits embedded in the surface of the flint. Phytoliths are microscopic platelets of silica which form the surface of cereal plant stems. We can, therefore, be fairly sure that these tools were used for harvesting. In contrast iron sickles, when they happen to survive, because of the nature of the material do not have such confirmatory deposits. Frequently their size and

3. Composite flint sickle. This sickle has been reconstructed to accommodate a series of flint flake blades in an antler handle.

appearance, while characterising the accepted sickle shape, are sufficiently different to question their efficacy as cereal-harvesting tools. Uncomfortably, the prehistoric cereal types are most efficiently harvested by picking the heads or spikes by hand, an edged tool being required for cutting the straw although this could just as easily be uprooted. Similarly plant gloss and phytoliths can be deposited on a flint from cutting grass. Thus the growing of cereals cannot be proved from the discovery of a sickle. Similarly milling or quern stones, most probably used to grind flour, can be used for a range of other processes and, therefore, do not provide conclusive evidence. These are examples of the inherent dangers of simplistic acceptance of form and shape. Similarly the typical stone axe, understood immediately as such and recognised as a tool for chopping wood and felling trees, could equally well be hafted differently and turned into a mattock hoe.

Nonetheless, caveats apart, a great range of tools and implements has been recovered from all periods. One of the most exciting discoveries has been the recognition of the stone ard tips from Orkney and Shetland indicating that ploughing with an

4. A group of stone ard tips recovered from an excavation in the Orkney Isles.

5. This reconstruction of a neolithic ard is conjectural but it is the only system which correlates with the wear marks on the stone ard tip and functions adequately in stirring the soil.

implement designed to stir the soil into a tilth can be assigned to the stone age. From subsequent periods wooden ards have been recovered which indicate at least three types of soil preparation. Work at Butser Ancient Farm has given great insight into all these ard types by direct practical experiment over many years. In addition a splendid survey of prehistoric agricultural tools has been published in this series by Sian Rees.

Physical evidence

In many parts of Britain evidence of prehistoric farming survives in the landscape. Traces of bronze age and iron age fields still abound, particularly in areas where modern intensive farming has not encroached. Most commonly these ancient fields are observable on the sides of hills and sloping ground where the rectangles, varying in size from less than a quarter of an acre (0.1 ha) to over an acre (0.4 ha), are marked out by lynchet boundaries. The lynchets are low banks which form through plough action and soil creep on the downhill side of the field. If another field is located immediately downslope the bank is accentuated by soil moving away from the upper edge of the field. In effect the lynchet has a positive element where the soil gathers and a negative element where the soil erodes away. Regularly these fields are found to be associated with trackways and settlements. Excavation of some lynchets has shown the presence of lines of stake-holes, indicating that some fields were fenced probably with interwoven saplings of hazel or willow. Conclusive evidence for hedges has not yet been recovered but it is most likely that thorn and hazel hedges were used. The evidence for hedges is most likely to come from analysis of snail evidence. Should there be recovered a significant proportion of shade-loving species of snails from excavated lynchet banks the probable existence of hedges would be implied. Much work still needs to be done in order to understand more completely the nature of field boundaries in prehistory.

Invariably the ancient fields which have survived in our landscape are to be found in areas of non-intensive modern arable agriculture. Primarily these are on the light upland soils of the chalklands. This has led to the dangerous assumption that prehistoric farmers were tied to these lighter soils because of the inadequacy of their tools and implements. Empirical trials have shown, however, that heavier soils can be cultivated with ards. Rather the surviving ancient fields demonstrate that historic and modern arable agriculture has taken place elsewhere. The major question is whether prehistoric farmers were restricted to lighter soils or whether ephemeral traces of their fields in other regions have been obliterated by subsequent farming activity. The presence of settlements from the bronze age and iron age period, now attested even on heavy clays like those in Northamptonshire, suggests the latter. Further, since prehistoric agriculture spanned some four thousand years it is inconceivable that during that time no attempts were made to explore different soils. If the attempt were made the results in increased yields would be totally persuasive that heavy soils were far better if more difficult than the shallow lighter soils.

Bog bodies

The recovery from peat bogs of human bodies like Lindow Man in 1984 and the celebrated Danish bog bodies found earlier in the twentieth century provide invaluable evidence for agriculture by virtue of the gut contents which have survived. The traces

6. On the southern slopes of Butser Hill in Hampshire a remarkable area of prehistoric fields has survived. The light snow cover and low lighting shows up the lynchet banks very clearly.

of their final meal, whether ritual or not, allow an insight into aspects of their diet and the preparation of the food. Similarly the recovery of palaeofaeces is important in this context.

Conclusion

The trace evidence that survives in all the different forms listed is, however, remarkably meagre. In terms of proving a case it is wholly inadequate and in the sense of modern forensic evidence it is barely circumstantial. The data extracted from a simple site excavation are rarely sufficient to give anything but the crudest understanding of that site. Our only way forward is the combination of many sets of data from different sites in order to begin to construct a broad overview. Thereafter the creation of hypotheses tested by practical experiment may provide more detailed boundaries of probability and point towards more sophisticated methods of investigation.

3
The sequence of development

The development of agriculture during the prehistoric period is the key to understanding how the landscape evolved. The earliest occupation of Britain was by the hunter-gatherer communities of the early and middle stone age, followed by what is called the agricultural revolution, which probably began in the late middle stone age and steadily gathered pace into the new stone age and beyond. Revolution is perhaps too strong a term to describe a process which lasted many hundreds of years but it is apt when taken in the context of a complete change in living patterns. The hunter-gatherer was totally dominated by his landscape and depended entirely upon his prey, whether plant or animal, for his survival. At this time the pollen evidence indicates that the landscape was primarily wooded with occasional glades; valleys were wet and marshy. The land was under the sway of the natural vegetation interrupted in its cycle only marginally by the animal kingdom. Necessarily there was considerable variety in the vegetation dictated by the nature of the soil and topography and critically by the climate. Some areas were more amenable than others but man's movements were always dictated by his prey, which in turn was affected by the seasons.

We have some evidence of this time of man's development and his movements from temporary camp sites like Starr Carr and Seamer Carr in Yorkshire, as well as the earliest evidence of man in north-west Europe at Swanscombe in Kent some 250,000 years ago. There is little doubt that conditions were hostile and that human communities were small and mobile. Observations of present-day hunter-gatherer groups allow some insight into both activities and group organisation. A hunting group can move only as quickly as its slowest member, a factor which would have controlled the number and age of children within the group. The greatest and most reliable source of food would probably have come from plants. Hunting and killing animals for food would have been a bonus. Modern parallels suggest that most of the work in food gathering was carried out by the women, successes at hunting being occasions for celebration rather than the norm. It would seem from the evidence available to us that this lifestyle was both successful and agreeable in that the work load was relatively light. Our problem is to explain why mankind abandoned this 'garden of Eden' to indulge in a work ethic which once begun allowed no backward glance. The actual introduction of farming was quite dramatic from two significant points of view. First, man changed his habitat by altering the natural state of the landscape, and second, because of the work load in a fixed location, increased his numbers to respond to the demand. One can only suppose that a period of ideal conditions gradually changed the community structure, a plethora of available food making the mobile way of life unnecessary. Increase in population led to expansion of land utilisation. A deterioration of climatic conditions would have led to food shortages requiring the first step towards agriculture. This step must have been the collection of food which could be stored against future needs.

It is at this point that the agricultural revolution must have begun. The range of storable foods is limited. Essentially these are hard dense seeds like those of certain grasses, plants like the bindweeds and nuts. The problem lies in the period of time the food must be stored. In the temperate zones for nine months of the year natural food supplies are minimal. Consequently the gatherer has to lay in stores and reserves to span this period at least. The revolutionary step was to cultivate deliberately the kind of plants which would provide the food in the necessary bulk to ensure survival. Undoubtedly this first step was a faltering one based upon observations of normal stands of certain plants and their basic requirements. Nut-bearing trees do not produce fruit until maturity, a process taking several years, nor do they necessarily bear fruit annually and therefore they would not be sensible plants to cultivate. Their use as occasional food sources would have continued even as they do today strictly as a gatherer resource. The first farmers had to select a plant which could be planted and which would mature in a single season. The evidence we have indicates that the plant chosen was wild einkorn *(Triticum monococcoides),* a primitive 'cereal' of the grass family. Its seeds are slightly larger than those of the average grass and are easy to store. When crushed into groats or ground into flour and mixed with water, the result is a nutritious, if to us an unappetising, food. Undoubtedly this recipe can be easily enhanced and probably was. Under cultivation the plant became domesticated, its seeds slightly increased both in number and size *(Triticum monococcum).* This process is the result of simple selection. Harvesting tends to gather the best seeds from the strongest plants. Replanting from this seed supply repeats the pattern and leads to gradual change.

The physical effect of this was dramatic. To grow plants

requires a prepared seed bed, the creation in effect of a field. To create such field plots, trees had to be felled. We have some remarkable evidence of this activity from Denmark. Pollen samples taken from peat deposits show a marked break in the record of tree pollens, which are replaced quite suddenly by grass and cereal pollens. Experiments were carried out by Steensberg and Iversen in an attempt to repeat the pattern in the Draved Forest. Trees were felled with stone axes, the brush and timber were burned to ash and cereals, barley and wheat, were planted in the cleared areas. Crop yields were measured but, significantly, the pollen evidence repeated the ancient pattern. This process is called 'landnam' or 'slash and burn'. In parts of the world slash and burn became the normal agricultural pattern. Clearings are created by slash and burn, crops grown until the land is exhausted, and then another area is slashed and burned in rotation. Given a large enough area to allow time for the trees to regenerate it is a viable system, but only in particular zones where tree regeneration will occur. Unfortunately soil erosion and depletion is the more usual successor to abandonment.

The labour investment in clearing land and growing crops, however, argues strongly for two specific results. The first result is the building of a permanent settlement and increased population to respond to the work requirement. The second result is the need not to exhaust the soil of its natural resources. In this connection experimental research at Broad Balk in the Lewes Agricultural Research Station at Harpenden, Hertfordshire, and the Butser Ancient Farm indicates that given this simple non-chemical agricultural system land exhaustion in Britain is a most unlikely outcome even if the same land area is used continuously for the cropping of cereals.

It would seem that once settlement had been achieved, the treadmill of agriculture was established. Man was at this point in direct conflict with his environment, his critical goal being domination. Failure clearly did not occur though details of the inevitable drama of the early phases are denied to us. Botanical chance, the wild cross of einkorn with a goat grass which gave rise to emmer wheat *(Triticum dicoccum),* which in turn further crossed to produce spelt wheat *(Triticum spelta),* aided these first farmers remarkably. Both emmer and spelt are heavy-eared bearded wheats with seeds similar in size to modern wheats. They are the predecessors of all wheats. The appearance of barley *(Hordeum)* also occurred in the neolithic though its history is extremely complex. Probably both the six-row and two-row forms come from one original ancestor, wild two-row barley *(Hordeum spontaneum),* which was initially cultivated like the einkorn. All the initial domestication of cereals took place in the Near East. For Britain the transition from hunter-gatherer to agriculturalist did not depend upon this exploitation of wild species producing in

7. This area of secondary woodland was used for a 'landscape' experiment at the Butser Ancient Farm. It is a typical mixture of hazel, thorn and ash trees within an abandoned coppiced woodland.

8. The experimental area within the woodland is shaded by trees and fenced against rabbits but not against deer. The stumps of the trees are left *in situ* within the cultivated zone. Despite four years of continuous cropping the pollen record was not altered in any way.

due course domesticates. These seeds were brought in as proven crops. This, however, in no way detracts from the problem posed by the temperate zones.

Settlement and the creation of a farm fixes man into a specific location. However, the real trial is not the creation but rather the maintenance of land control. The investment of labour is considerable. The direct result of this is an increase in family size and, therefore, in overall population, leading to more and more land being taken into cultivation to support the expanding population. The syndrome is clear. Similarly it is most unlikely that a settled farmer would readily move from one location in order to start afresh in a new one. Although we cannot prove that the first farmers did not shift about the countryside, the probability is low. The expansion of settlement through time across the landscape from the neolithic to the late iron age, a period of about four thousand years, culminates in total occupation of the landscape. Our evidence for the later periods is naturally far greater than for the earlier ones but the information we do have argues for rapid expansion.

A particular factor which necessarily boosted population was a significant climatic change that occurred sometime in the early centuries of the first millennium BC towards the end of the bronze age and beginning of the iron age. The weather changed from being relatively warm and dry to cool and wet. Although the change was not dramatic it was sufficiently significant to cause the abandonment of the region of Dartmoor for example. Throughout this region, which was heavily populated during the bronze age, farming became untenable. Similar effects can be seen in many highland areas where bronze age sites became marginal or inoperable, causing relocation of settlements usually to lower zones. The wet effect, however, was beneficial. A cooler, wetter climate specifically favours the growth of grasses, of which the cereals are a subtribe. Also, since the arable fields were inevitably infested with weeds as well as cereals, the increase in moisture enhanced the deterioration of fibre material within the soil structure, releasing more available nitrogen. In practical terms the farmer, whose technology in the form of implements and practice shows no appreciable change from the bronze age to the iron age, found his yields increasing steadily. The weather pattern today is argued to be similar to the late iron age and the reverse effect when the weather is dry and warm is lamented by modern farmers. This climatic change seems to have had a profound effect not only in terms of an expansion of population seen in the proliferation of sites, but also in the introduction of a new system of bulk grain storage in underground silos.

The sequence of farming development was not based only upon cereal cultivation, although this was clearly the most reliable source of food. The evidence suggests that farming was essentially mixed from the beginning. The maintenance of livestock adds a major extra component to the farming pattern especially in temperate regions. For Britain, where the snow cover is less significant and prolonged than on the continent, the problems are slightly less. Nonetheless, during the winter period when forage and grass are unavailable the farmer has to supply supplementary feed. Domestic livestock included sheep, goats, cattle and pigs. Whether the animals were kept for meat, milk, wool or draught purposes their upkeep during the winter was of paramount importance. The strange idea of mass slaughter of livestock in the autumn is palpable nonsense. Undoubtedly young animals not required for breeding would have been culled at this time but nothing more. The burden of providing fodder (hay, grass and probably dried leaves) must have been an important element of

9. Spelt (*Triticum spelta*), emmer (*Triticum dicoccum*) and einkorn (*Triticum monococcum*) wheat. Scale in centimetres.

10. The Moufflon sheep, the first domesticated sheep within the history of agriculture. (Photograph: A. Wyman.)

11. European wild boar (*Sus scrofa*).

the farming economy.

Like the cereals, the livestock also changed through the four thousand years of ancient farming. The bone evidence suggests that the earliest dometic sheep were Moufflons, followed by Soays through the bronze age and into the iron age. Gradually the bone evidence changes, indicating occasional four-horned sheep like the Hebridean and Manx Loghtan, giving way at the end of the iron age or beginning of the Roman period to the Shetland sheep. All these have survived as individual breeds to the modern day, usually, as their names imply, in remote areas and often in the feral state. Research at Butser and elsewhere shows, for example, that Soay sheep, which have survived on the St Kilda islands off the north-west coast of Scotland as feral animals with wool colours ranging from dark brown to oatmeal, may have been white-fleeced in the iron age.

Goat bones are remarkably difficult to differentiate from sheep unless either the skull or metatarsals survive, so the exact type of goat is almost impossible to identify today. Certain feral groups, the so called Old English Goat, are the most likely.

Cattle descend from the aurochs *(Bos primigenius).* In the neolithic the cattle were large but it seems that through the centuries a deliberate programme of selection led to a smaller compact beast suited for draught purposes. The modern equivalent is the long and medium-legged Dexter cattle.

Domestic pigs are all descended from one common ancestor, the wild boar *(Sus scrofa).* This is still a relatively common animal in Europe and regularly hunted in many countries, notably France and Germany. Undoubtedly the sport of hunting wild boar formed part of the ancient farming scene. However, capture and domestication of the piglets was an obvious and not difficult exercise. Management, on the other hand, is difficult to isolate. Given the two basic systems, herding or containment in a sty, the former is the most likely. Since the pig's major requirements are food and sleep they are relatively easy to control.

Although farming is essentially exerting control over the landscape, the final dominating factor is the landscape itself. The farmer has no option but to be in sympathy with his specific location with its microclimate. Thus, though mixed farming was the norm throughout prehistory, the weighting of the mixture of livestock to plantstock was subject to the soil, climate and topography of each area. This was as true in the neolithic as in the iron age and, indeed, as today.

4
Farming

Farming can be divided into three main elements of arable, pastoral and woodland management. Throughout prehistory all three elements would be incorporated into every holding but not necessarily in equal divisions. The landscape, its topography, soil and climate dictate which element predominates over the others. Woodland management, not obviously a principal agricultural activity, would have been a critical but not a dominant element from the agricultural revolution onwards. Timber was necessary for building construction and implement manufacture, fences and fuel.

Arable farming

Arable farming is all about the creation of fields, tilth preparation, manure, seed planting, crop management and harvesting. Provision of manure is a necessary cross link with the pastoral element. The primary objective is the production of sufficient storable food supplies for man and animal throughout the winter. Consequently food storage needs to be included in this section. It must always be stressed that we can never be sure that our views are correct and that any explanation is open to reinterpretation in the light of new archaeological discoveries. Inevitably our knowledge of the later periods of prehistory is greater than of the early periods. Nonetheless, given the initial slash and burn of the revolution itself we find stone artefacts which seem to be the shares of ards belonging to the neolithic period. The wear marks argue conclusively for their use in this way and experimental trials have demonstrated their efficacy. Even at this early stage in the development of farming man began to invent composite tools to ease the burden of work. Indeed, it was the easing of the work load rather than necessity which was the mother of invention.

For the mainstream of prehistoric farming, the bronze and iron ages, we have a variety of sources of evidence. From Scandinavia, southern France and northern Italy a number of rock carvings depicting agricultural scenes have survived. From these we can deduce the presence of three specific types of ard. The distinction between an ard and a plough is that the plough is fitted with a curved mouldboard which turns the soil over. The ard simply stirs the soil to a depth of about 30 cm (1 foot), its action being the

12. The rip ard or sod buster. This drawing is based upon one of many rock carvings from the Val Camonic in northern Italy. The scene it represents is extremely important for our understanding of prehistoric agriculture. The ard itself is simply an angled spike which is hauled through the soil. In practice it rips up the ground for about a couple of metres before it locks solidly into place and has to be lifted out before the operation is repeated. The main point of such an ard is most likely responsible for the ard marks found by archaeological excavation. The principal argument in favour of this explanation is the presence of the two individuals carrying mattock hoes. The one at the rear seems to be chopping at the ground, presumably breaking up the matted clods ripped up by the ard. The other person holding a hoe is leading the cattle. Perhaps his role is the releasing of the ard when it locks into place.

same as that of the modern scuffle plough. The first and most commonly depicted ard is the sod buster or rip ard used for bringing new ground into cultivation or recovering old fallow land. The second type, called a bow ard, is the normal cultivation ard which simply stirs the soil into a tilth, while the third type is a seed-furrow ard which forms a narrow drill for the seed in the prepared tilth. From the peat bogs of Denmark examples of the last two types have been recovered. In addition from a peat bog near Lochmaben in Scotland a main beam of a bow ard was discovered in 1870. All the ards were drawn by cattle.

This panoply of implements has considerable implications for

13. The Donneruplund ard. This reconstruction is based exactly upon the evidence provided by the original ard recovered from a peat bog in Denmark. It has been used in extensive ploughing experiments and has proved to be extremely efficient.

14. The Donneruplund ard in action being drawn by a yoked pair of Dexter cattle.

15. The heave of the ard. The soil is lifted by the ard and carried to swirl as it passes the foot of the main beam.

16. Representation of the Litsleby rock carving in Denmark (after P. V. Glob). The scene shows the drawing of seed drills with a specific type of ard. Two seed drills are depicted beneath the cattle and a third has just been commenced. The ploughman is carrying a bag, possibly of seed, in his right hand. His left hand guides the plough and holds a leafy bough. The purpose of this may well be a combination of fly swat and goad. The cattle appear to be yoked across the horns. The ritualistic nature of the carving is demonstrated by the phalluses on the cattle and the ploughman.

17. Reconstruction of the seed-furrow ard based directly upon the Danish bog finds of original ards from Hvorslev and Vebbestrup. The same type of ard figures in the rock carving from Litsleby. This reconstruction has been used in extensive trials at the Butser Ancient Farm. It produces an ideal seed drill in a prepared tilth.

arable agriculture. The sod buster or rip ard was probably responsible for the 'ard marks' which have been recognised on a large number of sites on differing soils ranging from light chalklands and sandy soils through to clay sites. It also argues that fallowing may have been an element of agricultural practice. The bow ard would have been the workhorse of all the ploughs. Its marks on the subsoil would, simply by the frequence of its use, have been self-cancelling. Experimental trials with this ard type have shown it to be splendidly efficient in creating a good tilth in a variety of different soils including the heavy clay soils of the midland region. The seed-furrow ard, however, is especially important. If its use was widespread then the crop management was advanced and sophisticated. If the seed were sown in drills, the crop would grow in distinct rows allowing inter-row hoeing of weeds. The presence of the hoe is well attested from the beginning, its original function of cultivation subsequently giving way to weeding.

18. Prehistoric plough marks carved into the upper chalk on Slonk Hill, near Brighton, East Sussex. These marks were probably created by the 'sod buster' ard. Note the truncated scores and irregular disturbances. (Scale in centimetres.)

The crops available to the bronze age and iron age farmer are identified by the carbonised seed remains recovered from occupation sites. The list is long and impressive. There were some ten cereals: emmer wheat *(Triticum dicoccum),* spelt wheat *(Triticum spelta),* club wheat *(Triticum compactum),* old bread wheat *(Triticum aestivum),* naked and hulled two-row and six-row barley *(Hordeum distichum* and *Hexastichum* var. *nudum),* rye *(Secale cereale)* and possibly oats *(Avena* sp.). Apart from the cereals there were the legumes, peas *(Pisum sativum)* and beans *(Vicia faba minor).* Given such a choice, the prehistoric farmer had considerable flexibility in how he utilised his land. The key to successful farming lies in an intimate knowledge of the soil and the microclimate of individual fields and even parts of fields. This knowledge is inevitably the result of experience. Some areas are better suited to barley, others to wheat. Similarly, one wonders if crop rotation was practised. The legumes fix nitrogen into the soil while cereals extract nitrogen. Cereals after legumes is the simplest rotation.

Aside from crop rotation the application of dung to fields was a major advance believed to have begun sometime during the bronze age. Clearly it is a critical line between the arable and pastoral elements of farming and a prime justification for mixed farming at any time. The evidence points towards the dung being collected in a farmyard midden along with domestic refuse and then transferred to the fields although the alternative of folding livestock on fields after harvest could also have been practised.

One further interesting aspect of the finds of carbonised seed has been the discovery of caches of pure seed of fat-hen *(Chenopodium album).* Today it is universally regarded as a useless weed. Once it was recognised as a valuable vegetable and was the precursor of spinach and cabbage. Its uses are considerable. The young leaves can indeed be used as a vegetable but in addition the whole plant can be used like a hay crop, sun-dried and stacked for livestock feed in the winter. Finally its seed can be harvested and ground up into flour and made into bread. There is, therefore, the possibility that it was a crop plant in prehistory. The last and potentially most valuable attribute of fat-hen is its normal germination period of June with fruiting in September. So, if all the normal crops were to have failed for some reason, fat-hen could have been a survivor and catch crop for the ancient farmer.

Apart from food crops there is ample evidence of flax *(Linum usitatissimum)* having been grown both for fibre from the plant

19. The Celtic bean (*Vicia faba minor*).

20. Woad (*Isatis tinctoria*), a first-year plant. This exotic may well have been grown specifically to manufacture a blue dye referred to by Caesar. The plant is a biennial, the leaves from the first-year growth being used for the production of the dye.

21. Woad in flower in its second year.

stems and for oil from the seeds. Similarly gold-of-pleasure *(Camelina sativa)* may have been grown for oil from its seed though it is regarded as a specific weed of flax fields. Other plants are evidenced in the archaeological data though many would have been garden rather than crop plants. Among these, for example, are the exotic woad *(Isatis tinctoria),* the source of the blue the British are reported to have tattooed themselves with (Caesar, *De Bello Gallico* books IV-V — *Britanni se vitro inficiunt),* and the opium poppy *(Papaver somniferum),* presumably for its medicinal qualities. There is some slight evidence too for the growing of hemp *(Cannabis sativa)* for fibre production as well as its other properties. Herodotus (book IV) gives details of how the Scythians enjoyed the vapour from hemp seeds cast on hot stones. There is little doubt that gardening in the sense of growing small quantities of a range of plants was practised from the neolithic onwards.

Harvesting and storage are the culmination of arable farming. The sickle is a common artefact throughout prehistory made from

22. A classic example of the evidence of an iron age four-post structure. The massive post-holes, nearly 50cm (20 inches) in diameter, form a rectangle approximately 2 metres by 2.5 metres (6 feet 6 inches by 8 feet). Their proximity to a group of storage pits suggests that they represent the foundation of an above-ground granary. Details of such a structure can only be conjectural.

23. The base of a simple haystack. The solitary upright pole around which the stack is built, perhaps accompanied by a circular dished depression, provides the only trace evidence the archaeologist is likely to find. The timber raft keeps the hay from direct contact with the ground.

stone and flint, bronze and iron in due turn. Exactly how harvesting was accomplished is difficult to determine precisely, especially for the cereals. It is not yet known whether crops were normally first deheaded, the straw being gathered as a second operation, throughout the prehistoric period or just during the iron age. Certainly both ears and straw were important harvests, the former as basic food, the latter for a variety of purposes ranging from cattle feed to roofing material. In the earliest period the product was most probably stored in containers like storage jars and sacks of leather. Rectangular patterns of four post-holes are thought to represent overhead granaries, strong wooden buildings raised off the ground on piles to protect the contents

24. Two completed haystacks with thatched roofs at the Butser Ancient Farm. Each contains approximately 1.5 tonnes of hay, just enough to maintain one cow during the winter.

from rodents and allow full air circulation. Solitary post-holes may well represent hay and straw stacks. By the iron age another means of storage appears, the underground silo or pit. These have been proved experimentally to be extremely successful storage units for both food and seed grains. Of most interest, however, is the capacity of such units, the average being at least 2 tonnes of grain committed to long-term storage. In effect, the storage facilities we can recognise underline the increasing success of the prehistoric farming system.

Pastoral

In the beginning man most probably relied upon plants rather than animals for his staple food supply but it is still a matter for debate which came first. The pastoral element of farming is almost inextricably entwined with the arable element and given mixed farming as the norm it is only the emphasis which changes. The real evidence for accurate interpretation of the arable side is virtually non-existent. The evidence provided by bone remains will always be partial and totally inadequate for estimating stock numbers and types. The very nature of archaeology which deals in, at best, hundreds of years rather than tens militates against comprehension of actual farming practice where most livestock complete their life cycle in less than a decade. Consequently any discussion of a pastoral element has to be the explanation of a hypothesis, the more so when one considers the great range of variables there are in stock management, as opposed to plant management.

The first domesticated animals seem to have been cattle and pigs, soon followed, as the landscape was changed from predominantly wooded to predominantly open land, by goats and sheep. Indeed, from the neolithic onwards in Britain bones of sheep and goats in greater or lesser proportions are found throughout.

25. The skull of a Celtic shorthorn (*Bos taurus*) recovered from the excavation of an iron age site at Bramdean in Hampshire.

26. The Dexter cow, the nearest modern equivalent to the Celtic shorthorn.

Occasionally bone analysis has allowed the conjecture of milking herds of cattle in the early period. Domestication, however, is a term we use rather loosely even today, describing as domestic dogs on the one hand, sheep and cattle on the other. In fact, except in special circumstances farm livestock are at best controlled feral groups. The control is exerted by restriction in the form of fences and regular food supplies, especially during the winter period.

We know from the rock carvings of the bronze age and iron age, and by the implication of ards even in the neolithic, that cattle were managed in two ways. The first was the real domestication system of training a pair of cattle to work together to pull ards and carts. Such animals would have received specialist treatment throughout their lives, probably being kept in the farmyard and never grazing freely. On the other hand milk and beef cattle (the distinction is totally artificial since the cattle would have been dual-purpose) would have been herded and grazed. Management is dictated first by the nature of the animal and second by the desired product. Milk, for example, can be obtained only from a cow which has calved, the normal lactation lasting only some ten months. For a regular supply, therefore, calving has to be organised. Whether prehistoric farmers shared the milk supply with the calf cannot be known. Similarly a cow does not mature until it is about two and a half to three years old but will then calve more or less annually for about five or six years. The problems of controlling the numbers of stock are obvious: how many calves to cull, how many to keep, bearing in mind the winter feed requirement of about 15 kg (33 pounds) of hay per day per animal during the winter and, critically, how many to keep to maintain the viability of the group.

The benefits of keeping cattle are meat, milk, leather and dung. The last is an important element for the arable fields. The average product of dung per cow is about 25 kg (55 pounds) per day and if the animals are kept overnight in a corral or byre the majority of this can be collected. The implications for the farmer are clear. The easiest hypothesis which is supported by archaeological evidence is the winter containment of cattle. For ease of winter feeding the animals are close to the stored food within the farm, and all the dung over the period of containment, usually reckoned to be some 120 days, is concentrated in one place. The traction cattle would be a further regular source, given their special treatment. The outstanding problem of cattle maintenance is an adequate water supply. The very basic daily requirement is 40 to 50 litres (9 to 11 gallons) per animal. There is, as yet, no archaeological evidence which throws light on how this was provided.

Cattle particularly raise the problem of grassland management in prehistory. The higher zones in hilly country and water-meadows are obvious alternatives for grazing but there had to be provision for hay and straw fodder for the winter. Tree-leaf fodder is a further probable component. Nonetheless considerable effort during the farming year had to be devoted to haymaking and leaf gathering to ensure adequate supplies. Also areas of grassland had to be set aside for haymaking, thus denying their use for grazing until well into July.

Sheep and goats, kept for wool, meat, milk and skins, similarly required far more management then the familiar idyllic picture of a small boy setting off with a flock into the hazy middle distance. Such pictures, usually borrowed from hotter climes, are totally inappropriate for temperate climates. Certainly by the iron age, with a high population density, the middle distance probably belonged to someone else! In winter both sheep and goats would

27. A leaf-drying rack. Bundles of ash and elm twigs are hung on the cross members to dry in the sun. The forked poles are used to raise the loaded cross members into place.

28. Soay sheep have identical bone structure to the sheep recovered from excavations of bronze age and iron age sites in southern England.

29. Shetland sheep. Towards the end of the iron age and through the Roman period the bone evidence for sheep is exactly similar to the Shetland breed. In addition sheep shears appear at this time. (Photograph: A. Wyman.)

need a modicum of supplementary feed for at least the same time as cattle. Goats, strictly Mediterranean animals, require more care and attention than the hardy sheep although their lactation after kidding can be maintained almost indefinitely. Sheep are perhaps the easiest stock to maintain but the old adage that a sheep is either fit and well or dead must have always obtained. If, as the monument evidence suggests, most of the land was covered with fields, one possible hypothesis for summer grazing is the use of such fields as paddocks with regular rotation of the flock from one to another. The benefits of this system are considerable. First, the field or paddock areas are evenly dunged, sheep dung being more beneficial than cow dung, thus enhancing the subsequent cereal crops. Second, because of the requirement to move the sheep regularly, they are subjected to handling and scrutiny. Last, given such a rotational paddock system, the grass and plant material is less lush and therefore less likely to carry an overburden of parasites, to which sheep are particularly susceptible.

Pig management represents similar difficulties except that their dietary habits are as adaptable as human ones. The simplest system to envisage is herding in woodland. Bearing in mind the critical role of woodlands in prehistoric farming, this would be an available resource and a useful way of keeping the shrubs and undergrowth at bay. At certain times of year, spring and autumn, the pigs would have been turned on to arable fields where their rooting activities would have loosened and broken up the soil, destroyed ground cover and provided the invaluable dung in exactly the right place. On the other hand there is no real evidence yet for pigs being kept in sties.

Woodland

Given modern agricultural practice, it may seem odd to include woodland as an element in agriculture. The most obvious link is a pastoral one. Both cattle and pigs, to a much lesser extent sheep and goats, are essentially woodland animals. The evidence from the first agriculturalists in the Near East suggests that cattle and pigs were the first animals to be domesticated simply because the first farms were carved out of woodland. Traditionally the pig is at home in woodland, especially liking marshy valleys. Behaviour patterns of pigs are much more like those of dogs and, indeed, humans in that they are extremely adaptable in both diet and habitat. Despite the ferocity of the wild boar, the piglets are easily tamed and quickly become dependent upon a farming

30. An old ash coppice. The trees here are ideal for building purposes, averaging between 15 and 20 cm (6 to 8 inches) in diameter. (Photograph: A. Wyman.)

organisation. There are two basic methods of keeping pigs, in a sty or in a herd. Evidence for the former is virtually non-existent and we must presume that an open grazing sytem within a woodland under the eye of a swineherd would have been the normal prehistoric system. Ideal woodlands comprise oak, beech, hazel, ash and thorn, in fact the typical woodland.

The second pastoral link is the provision of leaf fodder for livestock. There has been a long tradition in north-west Europe of collecting elm and ash leaves during the summer, drying and storing them exactly like hay ready for winter feeding to all kinds of livestock. The elm decline detected in the pollen diagrams has often been associated with the introduction of agriculture though disease is a more likely cause. Leaf fodder is a remarkably good winter feed and is preferred to hay by both goats and the prehistoric breeds of sheep. Observations of the eating habits of Soay sheep at Butser Ancient Farm have shown that they will browse leaves before they will eat grass, a habit in direct contrast to modern sheep breeds. The method of harvesting the leaves is to cut twigs and small branches, bundle them together and then hang them on a trellis rack to dry in the sun. Archaeologically the evidence for such a trellis rack is simply a pair of post-holes set about 3 metres (10 feet) apart in an open position, a not uncommon feature on hundreds of sites. Research is in progress attempting to prove the presence of leaf foddering by analysis of identified charcoal remains from bronze age and iron age sites. If the size and type frequencies are high for the typical tree types, particularly elm and ash, some confirmation of the practice might emerge.

Apart from the pastoral importance of woodland it is a critical element of agricultural economy on two further accounts. First, and perhaps less important, is fuel. Wood was the prime source of fuel for warmth and cooking. Of far greater importance was the requirement of building materials for both structures and fences. All the post-hole evidence for houses, other buildings and major structures indicates a preference for timber, usually oak, some 25 to 30 cm (10 to 12 inches) in diameter, straight and close-grained.

31. A typical section of hazel coppice backed by a belt of ash trees. The growth of the hazel rods is at the second-year stage. (Photograph: A. Wyman.)

32. A simple interwoven fence of stakes and hazel rods. This kind of fence has been evidenced by lines of stake-holes excavated along the boundaries of prehistoric fields.

Such trees average between forty-five and sixty years old and are the product of 'man-managed' woodland. In effect the trees have to be carefully maintained in controlled plantations to provide this kind of product. The large number of rural settlements and, from the end of the bronze age and through the iron age, the urban centres of hillforts, created a massive demand for such timber. The most remarkable aspect of this product, however, is that its provision spans at least two generations. Man was planning and maintaining a product which he would not use himself. The pattern was undoubtedly one of coppicing both oak and ash, a practice which persisted from the earliest times through to the twentieth century.

In contrast to building, timber was the basic requirement for fencing material. We have evidence for wattle hurdles from the neolithic settlements on the Somerset Levels. Lines of stake-holes found beneath lynchet banks indicate wattle fencing around arable fields. The raw material for such hurdle and fence construction comes from coppicing hazel shrubs. The hazel *(Corylus avellana)* is a robust and richly branched shrub which occurs over most of Europe and throughout Britain, especially on warm calcareous slopes. Traditionally it has been coppiced for its

33. Reconstruction of an original iron age so-called sickle. It is, however, most unlikely to be a sickle but rather a tool for splitting hazel rods, a purpose for which it is admirably designed.

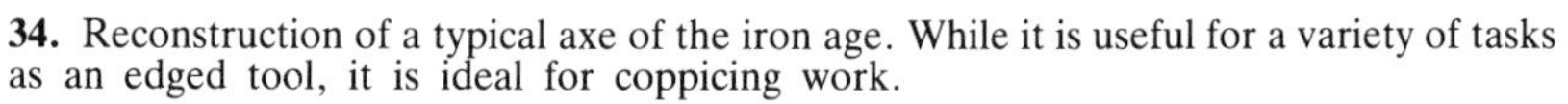

34. Reconstruction of a typical axe of the iron age. While it is useful for a variety of tasks as an edged tool, it is ideal for coppicing work.

supply of stems for manufacture into sheep hurdles. All the evidence points to it having been coppiced on a grand scale throughout prehistory. For example, at the Demonstration Area of Butser Ancient Farm, a small site of some 2.5 ha (6 acres), over 500 tonnes of hazel rods are used in standing fences. There is also abundant evidence from all periods that hazel nuts were harvested as a food source. In areas where hazel is not abundant, willow would have been used for fencing instead. Osiers for basket construction represent another important woodland product.

In the later stages of prehistory, the metal ages, a major product of the woodland must have been charcoal. It is extremely difficult to isolate production sites simply because charcoal burning is a woodland industry. However, the requirement for charcoal must have accelerated through time. An alternative source of raw material for charcoal production could well have come from old fences. The life span of an interwoven hurdle type is limited to about six years and it is most unlikely that such an amount of wood would have been wasted or even been reserved for domestic fuel. In this case charcoal production could well have come within the compass of a farm activity.

Conclusion

Farming does not just happen, whether it is ancient, historic or modern. It is an extremely complex process requiring great skill in balancing the different component elements, inputs and outputs, gambling against the greatest uncertainty of all, the climate, and managing to have sufficient reserves to survive the worst possible outcome. This hypothesising about the remote past may seem over-sophisticated but when the options implied by the archaeological data are all taken into account it may be yet too simplistic.

5
The farming year

In this chapter the farming year has been set out almost as a work list through the seasons. Planning is an integral part of farming simply because if things are not done in season the dependent sequence of events will not occur. Much of the work in agriculture does not necessarily have a visual impact and, therefore, to the casual observer, either did not happen or did not exist. The fact that such work is often intensive and hard consequently goes unrecognised. Similarly, since popular perception of farming is limited to the achievement peaks like lambing, haymaking and the harvest, the pattern of preceding events needs to be stressed. Thus the archaeological data, the nature of the evidence we have for prehistoric farming, demands specific work schedules, event patterns and seasonal rhythms. In other words, to produce the artefacts, carbonised seed, pollens, ard marks and ards, animal bones and bodies, various processes had to have happened.

Autumn

The autumn season traditionally begins the farming year. At this time the work schedules are geared to the harvest of the following summer. To a large extent the work done now will ensure success or failure at that time. Manuring, autumn ploughing and seed planting are the major activities. By the mid bronze age there is evidence of manuring of fields and, indeed, for the planting of autumn-sown cereals. Experimental research has demonstrated that both emmer and spelt wheats as well as the barleys can be planted successfully in the autumn. There are two particular advantages from autumn sowing: first, the pruning action of the winter frosts causes the wheat to tiller more abundantly in the following spring and give a heavier harvest; and second, because the autumn-sown crop matures earlier than the spring-sown, the harvesting time is staggered, allowing more land to be planted and harvested successfully. Inevitably the gamble with an autumn-sown crop is greater since in exceptionally hard winters with minimum snow cover the frost can destroy the crops.

Therefore the fields selected for the autumn-sown crops, most probably those away from the high plateau areas and the valley bottoms where frost action is most severe, would be first manured. Either cattle or sheep would be folded on the field to eat off the autumn growth of weeds and grasses and evenly dung the area, or the manure would be carted from the farm and spread manually. It is important, once the manure is spread, to plough as soon as possible. Manure spread thinly and evenly is highly beneficial, but in concentrations it will 'burn' the ground, making it infertile for several seasons to come. Ploughing and cross-ploughing with the beam ard at the rate of under half a hectare (about an acre) a day to break up the soil thoroughly would follow. At this time, too, hand weeding out of couch grass, that inevitable and pernicious weed of cereal crops, would have been vital. Couch grass is impossible to eradicate but vital to control. Furrows (the beam ard creates a furrow a good 30cm (1 foot) from crest to trough) were levelled to an even tilth. How this was done is uncertain but traditionally a heavy log dragged across the furrows was used. Finally came the seed-drill ard and the planting of the seed.

Once the fields were planted the focus shifted back to the farmstead and the preparations for the long dark days of winter. Sheep particularly need attention at this time. Culling the

35. The furrows created by the Donneruplund ard average some 30 cm (1 foot) from crest to trough. The soil structure is thoroughly stirred although there are no traces of ard marks in the underlying rock.

non-breeding stock, principally the young ram lambs and old ewes, came first. There is no point in using valuable winter feed on these animals. This activity must in turn have put pressure on the domestic economy with butchery, drying, smoking and salting of meat. The rest of the flock would require attention to their feet, the hooves needing to be trimmed back at least twice a year, but especially now with the hard ground in prospect. Culling decisions, too, would have to be taken with all the other livestock. The animals to be overwintered would be grazed on whatever grass was still available, the anxiety always being to preserve the winter feed supplies for as long as possible. Goats must have presented a major problem. Their predilection is for warm dry weather and the cold can kill them. Their maintenance during winter must have required the provision of housing of some kind which had to be prepared at this time.

In practical terms in the autumn there was a myriad of small jobs of preparation, tiny in themselves but in sum a daunting work programme.

Winter

To the layman the winter is a time to be indoors, keep warm and await the spring. For the farmer of antiquity it was a time of intense activity. The ploughing continued right up to the end of the year with a concentration upon the rip ard breaking open old fallow or creating completely new fields. The objective was to rip open the ground to allow the frost to help break down the soil before the spring cultivation. The focus in winter shifts to the woodland. Once the leaves have fallen and the sap has stopped rising coppicing begins. The requirement for wood on an ancient farm was large and ongoing. Hazel and willow for fencing were vital. Each year saw need for repairs, refurbishment and replacement of fences. Many tons had to be cut, bundled and transported from woodland to farm. Fence posts, young ash and oak trees, had to be selected, cut and brought in. With a fence post roughly every metre along the length of a fence, many hundred would have been needed. Building timber, too, had to be felled, trimmed and hauled. Doubtless nothing was wasted. Wood not required for specific purposes would have been collected for fuel, probably not for the present winter but for the following one.

One further interesting possibility was the search for useful timber shapes. Many of the implements, especially the ards, use naturally curved timbers for the main beams and handles or stilts.

Although the shapes used certainly grew that way, it is most probable that branches and even tree trunks were trained into the desired form and shape for future use.

Winter also brought the need for supplementary feeding of livestock, including the daily provision of water. This activity, for which there is the minimum of archaeological evidence, nonetheless must have taken much time and great effort. The daily requirement of a single cow averages some 75 litres (16½ gallons) of water. The prehistoric sheep breeds normally require little water provision, gaining adequate moisture from the grass and herbage, but when they were fed on dry feed and in freezing conditions water had to be made available.

Finally the winter is the ideal time for general repairs both to buildings and implements. There is never enough time or opportunity during the other seasons and chances lost now are rarely made up.

The most important activity of the winter apart from human and livestock maintenance, although this is arguably the prime purpose of farming in the first place, is the preparation of seed for the following spring. The prehistoric cereals of emmer and spelt particularly have to be broken down from the reaped ear or spike at least to spikelet form if not to the naked seed. This is a hard and time-consuming job.

Spring

The advent of spring is marked by lengthening days, less frequent frosts and warmer weather. The work begins fairly slowly as the farmer waits for the conditions to become favourable. Such opportunities vary from day to day but as they occur each has to be capitalised upon. The major concern again is the ploughing of the fields, creating seed beds and planting the seed, first the cereals and later the more frost-susceptible legumes, the peas and beans. Last to be planted would have been the flax fields; the least hint of frost on the young seedlings is enough to destroy the crop.

At this time too the manure is cleared from the byres and middens, loaded on to carts and taken to the fields, where it is spread evenly across the soil. This task always precedes the ploughing itself so that the manure is mixed well into the soil. Such a process is easy to dismiss without realising the nature of the quantities and weights involved. Many tens of tons are needed for manuring to be successful.

In the latter part of spring, late April and early May, the

36. Emmer wheat (*Triticum dicoccum*) infested in late spring with the worst weed of cereal crops, charlock (*Sinapis arvensis*).

lambing begins. The prehistoric breeds rarely have lambing problems but there are always exceptions. Rejected lambs would have provided problems and one wonders if the prehistoric shepherd accepted these as losses or attempted to raise them on a bottle equivalent. Also, given the concern and undoubted skill of ancient farmers in cattle management, it is probable that cows were put to the bull around midsummer so that calving would take place the following spring when grass was plentiful. This would suggest that bulls were kept separate from the other cattle for the majority of the year, a system which would further complicate the handling problems.

One of the most tedious and time-consuming jobs of all on the farm, hoeing the crops, is most necessary in the late spring. From the types of plough we have evidence for it seems clear that seed was sown in rows, probably set about 30 cm (1 foot) apart, to allow room for the hoers to do their work. Unless the weeds are kept down they can swamp a crop completely. Of all weeds at the beginning of the season the charlock *(Sinapis arvensis)* is the greatest enemy of the farmer. Virtually impossible to eradicate, it has to be kept at bay. It usually needs at least three hoeing sessions to ensure that the emergent crop will flourish.

Summer

Of all the seasons the summer has the finest weather. In this season the farmer sees the culmination of his hard work and planning. At summer's end he will be able to realise the results of the season's gamble. In practical terms it is perhaps the hardest season of all and not until the harvest is safely gathered will there be any relaxation.

The first major task of summer, apart from the normal grazing routines for livestock, is the haymaking. It is extremely difficult to prove that hay was made or when the practice started but suitable tools were available from the neolithic onwards and hay was a critical winter fodder without which it would be virtually impossible to maintain any livestock successfully. One possible proof for hay and even straw stacks is the regular presence on prehistoric sites of solitary post-holes, occasionally surrounded by a circular shallow gully about 3 metres (10 feet) in diameter. Traditional haystacks in Britain and other European countries used to be round with thatched roofs, the stack being built around a solitary upright pole to give it vertical strength. The basic ambition is the making and storing of enough hay for all the livestock to span a minimum 120 days of winter with a safety margin. The prehistoric farmer no doubt had to calculate his requirements exactly like his modern counterpart with broadly the same criteria. Cattle need approximately 15 kg (33 pounds) of hay per day, sheep and goats about 4 kg (9 pounds). There are several hypotheses as to how land was managed for hay production. Water-meadows would have been ideal since, because of flooding, such areas would have been too unpredictable for crop growing. The fallow of arable areas is an alternative option, while a third is the exploitation of maintained grassland, taken out of the grazing regime from the spring until late summer. The feed supplementary to hay may well have been tree-leaf fodder. Collection of ash and elm leaves could have been carried out throughout the late summer; they would have been dried on racks and then stored either in stacks or in buildings.

Almost simultaneous with haymaking is the shearing, or in the case of the Soay, the plucking of sheep. The wool of the Soay sheep is a short staple which they shed naturally through the month of June. The difficulty is in selecting the right moment to pluck their wool when it is just ready and before they rub it off. Since wool was one of the primary reasons for keeping sheep this timing would have received careful attention. In the later iron age, when the breeds developed, often the wool was sheared in the normal way. Examples of sheep shears have been recovered from this period.

The harvesting of the crops is spread across a two-month period, especially if the planting is programmed correctly. First the autumn-sown cereals are gathered in, followed by the spring-sown. The exact method of harvesting can only be conjectured but whatever method was used urgency was essential. When the crops were ready and the weather suitable the only priority was the gathering of the harvest in the shortest possible time. All the crop, both seed and straw, was important, the seed primarily for human consumption, although goats and sheep can eat whole cereals, in contrast to cattle, for which whole cereals are dangerous and can easily be lethal. For livestock the straw provides another winter feed alternative. Unlike modern wheats, the straw of both emmer and spelt is palatable to cattle. Straw had many other uses, especially as thatch. The storage of the harvest presented problems. In the earliest periods it was most probably stored in the houses. Later specialist above-ground granaries set on four posts, allowing free air flow around and under them, were built. Another system occurs in the iron age

37. Emmer wheat (*Triticum dicoccum*) just before harvest.

along with the above-ground granaries, this being the underground pit or silo. Much research has been carried out into pit storage of grain at Butser Ancient Farm, proving that this system of anaerobic storage was extremely successful. However, before storage the ears of the prehistoric-type cereals had to be thoroughly broken down into spikelet form without the awns. If this is not done, too much air remains in the pit and causes collapse of the seal and consequent failure.

After the cereals come the legumes, the beans and peas. The farmer undoubtedly enjoyed a proportion of the crop as green vegetables but the majority would be allowed to ripen and dry on the plant. Once the seeds were hard the crop would be harvested and stored. Simple soaking in water overnight before cooking makes both peas and beans a valuable food resource.

38. Sheaves of emmer wheat straw harvested from the field and transported into the farm. Straw was used for many purposes including livestock feed and thatching. The ears of the wheat were harvested separately before the straw.

39. The thatcher's stack. The sheaves of straw are laid between two upright posts and weighed down by a horizontal timber. The straw is carefully drawn from this stack and arranged into the clean bundles called yealms for the thatcher to use.

40. A grain storage pit. During the storage period the grains adjacent to the seal and the pit walls germinate, releasing carbon dioxide gas as a waste product and forming a protective skin. The carbon dioxide preserves the bulk of the grain and destroys the germinated seeds. The low winter temperatures inhibit the activity of fungi and bacteria. The germinability of the stored grain is unaffected.

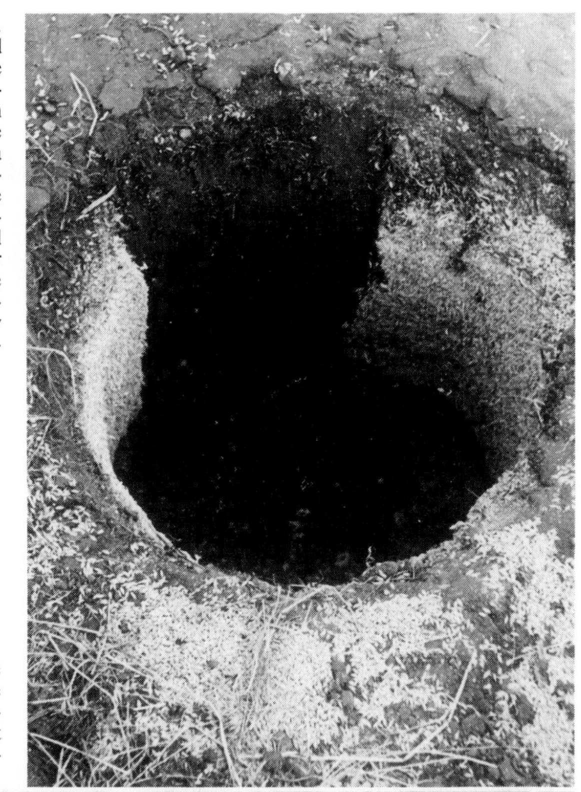

41. A typical saddle quern of the late bronze age or early iron age recovered from a shallow pit on Little Butser, Hampshire.

Flax perhaps completed the principal crops. Its harvesting is but a stage in its treatment. The uprooted plant would be threshed, its seeds collected to be crushed for their oil, and the plant stems themselves laid out in a damp place to be retted and subsequently combed of the fibres to be manufactured into linen threads.

Alongside the formal harvest the wild or natural harvest would not have been ignored. Fruits, berries and nuts are obvious choices, less so perhaps the collection of reed *(Phragmites* sp.) for thatching and bedding purposes, and maybe the long strands of old man's beard *(Clematis vitalba)* for binding, and nettles for thread. The potential of the landscape was without doubt fully exploited.

This seasonal task list presents just the bare bones of what had to occur for farming to have happened at all. Many of the lesser tasks have been omitted through constraints of space. Nonetheless from the neolithic revolution when farming began through to the end of the iron age the pattern of mixed agriculture was such that for it to succeed, and it did suceed, this work programme through the seasons was inevitable.

42. A rotary or oscillatory quern of the later iron age recovered from Maiden Castle in Dorset. It is possible to grind a fine flour from such a quern within a very short time. (Photograph: A. Wyman.)

6

Conclusion

Man is inevitably and inexorably moulded by his landscape. The farmer from the earliest times has sought to stamp his control over the landscape but can succeed only in those activities which are sympathetic to the location, its microclimate, its soil type and its topography. While the changes that have taken place may seem total and dramatic, nonetheless they have to be within the potential of that region. Until the present generation this has been wholly true. It is undeniable that man's stupidity and greed have destroyed regions of the world, created deserts and dust bowls, felled forests and drained swamps. Today, however, setting aside all dramatic issues of so-called global significance, the nature of farming is completely changed. There has been an agrochemical revolution so successful that it has occurred almost without notice by the majority of the population. After six thousand years of continuous agriculture in Britain, the farmer can at last dictate. It is still a gamble, if only because of the vagaries of the climate, but that gamble is charged with less chance than ever before. The children of this generation, essentially the products of urban environments, will have a perception of agriculture completely different from that of any previous generation.

7
Further reading

Bowen H. C. *Ancient Fields.* British Association for the Advancement of Science, 1961.

Clutton-Brock, J. *Domesticated Animals from Early Times.* British Museum, 1981.

Coles, J. M., and Orme, B. J. *Prehistory of the Somerset Levels.* Somerset Levels Project, Cambridge and Exeter, 1980.

Curwen, E. C. *Plough and Pasture.* New York, 1961.

Fowler, P. J. *The Farming of Prehistoric Britain.* Cambridge University Press, 1983.

Glob, P. V. *Ard og Plov i Nordens Oltid* (with English summary). Aarhus, 1951.

Mercer, R. (editor). *Farming Practice in British Prehistory.* Edinburgh, 1981.

Rees, S. E. *Ancient Agricultural Implements.* Shire Publications, 1981.

Renfrew, J. M. *Palaeoethnobotany.* Methen, 1973.

Reynolds, P. J. *Farming in the Iron Age.* Cambridge University Press, 1976.

Reynolds, P. J. *Iron Age Farm : The Butser Experiment.* British Museum, 1979.

Ryder, M. L. *Sheep and Man.* Duckworth, 1983.

Stead, I. M., Bowke, J. B., and Brothwell, D. *Lindow Man : The Body in the Bog.* British Museum, 1986.

Steensberg, A. *Draved.* Copenhagen, 1979.

Taylor, C. *Fields in the English Landscape.* London, 1975.

8
Places to visit

Andover Museum and Art Gallery, 6 Church Close, Andover, Hampshire SP10 1DP. Telephone: Andover (0264) 66283. Includes the Museum of the Iron Age in Southern Britain.

British Museum, Great Russell Street, London WC1B 3DG. Telephone: 01-636 1555.

Butser Ancient Farm Demonstration Area, Queen Elizabeth Country Park, near Petersfield, Hampshire. Telephone: Horndean (0705) 598838.

Cotswold Farm Park, Guiting Power, Cheltenham, Gloucestershire. Telephone: Guiting Power (045 15) 306 or 307.

Ancient fields can be seen as field monuments in many parts of the British landscape. Look especially in Hampshire, Dorset, Wiltshire, Sussex, Berkshire, Yorkshire, Northumberland and south-east Scotland.

Most museums with a prehistoric section have displays of agricultural implements from all periods.

Index